DEDICATION

This booklet is dedicated to the individuals living with dementia who made this booklet possible in order to support and enlighten the way for others newly learning they are living with dementia.

ACKNOWLEDGMENTS

For their personal contributions, we gratefully thank Terry Berry, Robert Bowles, Christine Bryden, Ellie Clark, Belva Dickman, Michael Ellenbogen, Paulan Gordon, Alexander (Sandy) Halperin, Truthful Loving Kindness, Brian LeBlanc, Carole Muliken, Mary Radnowsky, Laurie Scherrer, Kate Swaffer, Teresa Webb, and John Wood.

With gratitude to Karen Love who guided the development of this booklet from concept through publication with unwavering dedication and inspiration.

Dear Friend:

We are living with symptoms of dementia and understand the complex feelings and challenges experienced by people living with dementia symptoms. We found that most of the information currently available is focused on cognitive changes and impairments and does not provide insights and help with learning how to LIVE with a long-term degenerative health condition.

We've compiled our insights and experiences in this booklet to serve as a guide and a support for you. May the booklet be a useful source of enlightenment, hope and inspiration, and help you know that you are not alone.

The **Resources** section in the back provides information about connecting with others living with dementia as well as information on how to access additional helpful resources.

Caring wishes,

Dementia Action Alliance Advisory Council Members & Partners Living with Symptoms of Dementia

> *"I'm still me with kinks."* — Joan[1]

---

[1] Banks,S., Medine, J., & Morhardt, D. (2008). *What Happens Next: A booklet about being diagnosed with Alzheimer's disease or a related disorder created for you by others with a diagnosis of dementia.*

Table of Contents

About This Booklet 4
Learning You Have Dementia 5
Grieving is Normal and Natural 10
Telling Others You Have Dementia 13
Common Misperceptions about Dementia 15
What You Can Do about Misperceptions 18
Well-Being: A Newfound Friend 20
Keep Your Spirits Up 22
Develop and Maintain a Caring Support Network 25
Limit Stressful Experiences 27
Seek Fun and Interesting Things to Do 29
Add Lots of Laughter to Your Life 31
Increase Opportunities to Fill Your Spiritual Soul 33
Be Open-Minded to Creative Strategies 35
Eat and Drink Healthily 38
Learn Good Sleeping Habits If You Don't Have Them Already . 41
Be Physically Active 43
Finding Silver Linings in Your Unexpected Journey 45

Attachment I – What is Dementia? 46
Attachment II – Assessment and Diagnosis of Dementia 48
Attachment III – Your Brain 50
Attachment IV – Managing Dementia Symptoms: Medications 54
Resources 56
About the Dementia Action Alliance 57
The Artwork 58

About this Booklet

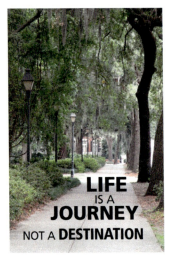

Throughout life people are on journeys — the journey to grow up, get educated, have relationships, find one's way in the world, and growing older among many other life transitions. Life takes an unexpected turn when you learn you have dementia. It is a transition in the road of life, however, not a stop. There is still much life to live and enjoy. This booklet provides insights and information about the journey of LIVING life with dementia from individuals who are living with dementia symptoms. The writing is facilitated by staff at the Dementia Action Alliance.

This booklet is for people who are living with early to moderate dementia symptoms. There is an abundance of information available publicly and in the marketplace about cognitive and functional changes that can occur with dementia, but there is very little information about LIVING life with dementia. Most people live for many years with dementia symptoms. Public media has led people to believe living with dementia is fraught only with terrible experiences. Not so! Baby boomers and younger generations of people currently living with dementia are speaking out and explaining that their lives haven't ended. Life can still be full of fulfillment, happiness, fun, and intrigue as well as challenges, fears and disappointments.

Please visit the **Resources** section at the back of the booklet for many other sources of helpful information.

Learning You Have Dementia

For some people, receiving a diagnosis of a form of dementia, including Alzheimer's disease, Lewy Body, vascular, fronto-temporal, or mixed dementia, can be a relief to learn that there is an explanation for the cognitive symptoms being experienced. For others, learning they have dementia is traumatic. Many people describe the post-diagnostic time as a time of withdrawal and feeling angry and/or depressed.

> "Learning you have dementia is like receiving an unseen right hook to the chin! You fall to the mat and hear the referee start the count...1...2...3...
>
> You try to get up but you cannot move, you're numb... 4... 5...
>
> You shake your head trying to get a bit of clarity...6...7...8...
>
> The numbness starts to fade...9...
>
> You hear a voice in your head screaming at you to 'GET UP! DON'T LET THIS BEAT YOU!' and you find a strength you never knew you had and rise to your feet before the 10 count."
>
> — Brian LeBlanc

Brian LeBlanc's 10 count lasted two days after he was diagnosed with early onset Alzheimer's disease at the age of 54. Curled up in a ball on his bed, Brian remembered his grandfather, his father, his wife's grandmother and his mother all having dementia and envisioned them with advanced symptoms. Then he thought about his wife, the love of his life, and two children and decided there was no way he was going to be counted out. So, each and every day when Brian wakes up, he pledges to make a difference by speaking out.

Christine Bryden explains her reaction as follows —

> *"The day before my diagnosis, I was a busy and successful divorced mother of three girls, with a high-level job with the Australian government. The day after, I was a label: Person with Dementia. No one knew what to say, what to expect of me, how to talk to me and whether to even visit me. I had become a labeled person, defined by my disease overnight. It was as if I had a target painted on my forehead, shouting out for all the world to see that I was blindfolded, no longer able to function in society.*
>
> *My first two years (post-diagnosis) were a struggle of living a life transformed by this label of dementia. I felt shame and retreated from society."* (Bryden, 2016, p. 54[2])

---

[2] Bryden, C. (2016). *Nothing about us, without us! 20 Years of dementia advocacy.* Jessica Kingsley: London.

Christine was encouraged to write about her feelings and ended up writing a book that helped to transform her despair into activism. Brian rebounded in two days; Christine's process took two years. There is no right amount of time — everyone is different. Brian is innately oriented to positivity which helped buoy him. An attitude of positivity is the goal for its therapeutic benefits. Instead of wallowing in negative thoughts and feelings, a positive attitude helps you stay directed to constructive thoughts and feelings. It's natural to feel down at times. Brian lets himself occasionally have a pity party. Since it's a party of one, he becomes bored quickly and moves on to doing something productive.

For some, how the diagnosis of dementia was delivered adds to the trauma of the experience. "You have about five years till you are really demented, then about three years after that in care before you die," was what Christine Bryden was told by her doctor when diagnosed with Alzheimer's disease. Kate Swaffer was told something similar and counseled to put her end-of-life affairs in order by her doctor when she was diagnosed. The actual words may

Another Time, Another Place 1
Photography, 2015 • Christopher Schneider

vary a bit, but all too often these are the discouraging statements people and their family members receiving a diagnosis hear. Let's add this to the list of things about living with dementia that need to change! There is still a lot of life to live and joy and love to experience.

Some people bury themselves in denial. The problem with denial is it only extends the inevitable instead of dealing with it. "Denial pushed me to maintain a façade of normalcy, which then tended to increase the denial. Occasionally I'd get hysterical out of desperation or frustration, which made my problem look psychiatric rather than neurological. Thank God for a therapist familiar with dementia, for a few old friends and for Internet buddies" (Bryden, 2016, p. 23[3]).

Richard Taylor, a giant in the dementia self-advocacy world, was diagnosed with Alzheimer's disease at the age of 58. He bravely spoke out about living with dementia across the world when it was taboo to do so.

> "I would like to be fondly remembered as Richard, the tall guy with the beard who lived for a still as yet to be determined number of years as a giving, challenging, loving and engaging human being. I am Richard who seeks to live a purposeful and purpose filled life up to and through my last breath."

---

[3] Ibid.

Richard died in July, 2015 of cancer. He did live a purposeful and purpose filled life up to and through his last breath. His wise words of advice live on —

> "Here is what I wish someone had told me about three weeks after I was diagnosed. You are not now moving faster towards the day of your death than you were the day before being diagnosed. Dementia is not a death sentence. It is a wakeup call to live your life, today and every today for the rest of your life as fully as possible…You are not alone in this journey unless you chose not to reach out to seek out others who share similar disabilities with you… You are not fading away, you are changing. But, of course, we all are changing all the time. We don't always have a choice of what will happen to us, but we always have a choice on how we respond… Live your life to its fullest. Don't try to live without or around your symptoms. Live with them, they are a part of you. Speak out. You owe it to yourself and others. If we don't tell others what it's really like to live with the symptoms of dementia how will they ever know?" (Facebook, richardtaylorphd)

The term 'dementia' is used throughout the booklet. Like cancer, dementia is an umbrella term that includes many different forms of dementia, such as Alzheimer's, Lewy Body, frontotemporal, vascular and mixed dementia.

Grieving is Normal and Natural

This information about grieving is not intended to be maudlin, but rather to address a normal human reaction to learning you have dementia. You may experience a wide range of emotions including numbness, denial, sadness, guilt, anger, anxiety, helplessness, frustration, and blame. These symptoms of grieving are a normal and natural response to loss — in this case the loss of some cognitive functions and abilities. Grief is an individualized process with no set time frame as it is unique for each individual. It is important to allow yourself to experience grieving in order to move beyond it.

Knowing what to expect along the grieving process can be helpful information. Dr. Elisabeth Kubler-Ross's famous stages of grief are a useful general guide[4]. The stages are not linear or predictable, and may occur in any order if at all.

- Denial — Being unable or unwilling to accept having dementia. It may feel as though you are experiencing a bad dream and are waiting to "wake up" so things will be back to normal.

- Anger — Feeling angry about having dementia and the unfairness of it.

- Bargaining — Pleading to a higher power to undo the condition with promises such as better behavior or making significant life changes in exchange for reversing the condition.

---

[4] Retrieved online on November 2, 2016, http://www.ekrfoundation.org/five-stages-of-grief/.

- Depression — Feeling a range of emotions and behaviors such as sadness, irritability, guilt, sleep too little or too much, change in eating habits, withdrawal from people and activities.
- Acceptance — Accepting the condition, you are able to re-engage in daily life.

This is a time to be especially gentle and patient with yourself. You may want to confide in one or more trusted friends about how you are feeling so that your emotions don't bottle up inside you. Seeking professional support, such as through a therapist or psychologist, is another helpful option. There are now a number of online dementia mentor and support groups run by people with dementia in early to mid-stages who are going through the experience of living with dementia and can provide invaluable support. See the back of the booklet for a list of resources.

Dementia is generally a progressive condition meaning that over time brain changes progress. When new impairments surface this may cause you to again experience some grieving symptoms. This is normal and OK. It's helpful to know, though, that feelings of grief may appear from time to time.

Suggestions

- Be kind to yourself. Do what makes YOU feel good!
- Confide in trusted family or friends. There is truth in the well-known anonymous quote — "A sorrow shared is a sorrow halved."
- Do things that fill your soul such as nature walks, spending time with people dear and near to you, and going to a place that has special meaning for you.

- Speak to a Dementia Mentor, an individual living with dementia, who has been on the journey and can provide helpful and practical insights and information — https://www.dementiamentors.org/mentor-sessions.html .

- Participate in an online café and talk with others living with dementia — http://www.dementiaallianceinternational.org/events/cafe-le-brain/.

Leaves
Acrylic, 2016 • Linda Birtles

Telling Others You Have Dementia

Because of the stigma associated with having dementia, many people are not comfortable telling others they have it. The downside is that the silence adds to the closeted nature of living with dementia. Fortunately, there is a cultural shift underway as baby boomers and younger generations of people impacted by dementia are speaking out. Unlike previous generations, they are more open and vocal about everything — including talking about their personal experiences of living with dementia. Speaking and writing publicly about living with dementia is helping to expand knowledge of and understanding about the condition which, in turn, is helping decrease the stigma associated with it.

Comfort in confiding with others about having dementia aligns with a person's personality type and age. Some people are naturally open and comfortable sharing information freely while others are more guarded and tell only immediate family members and close friends. If you are currently employed, there may be reasons to be circumspect about disclosing your condition. Speak with your physician and possibly a family law attorney for clear advice.

One person's therapist told her to tell no one; that her friends would fall away. She decided not to follow the therapist's advice, but did wait six months before telling her children. She wanted time to get used to changes in her life, and to be able to talk with her children without becoming overly emotional. The best advice is to follow your heart AND confide in someone close to you before sharing any information further. This gives you a chance to say what's in your heart. After you can decide what you want to say to others.

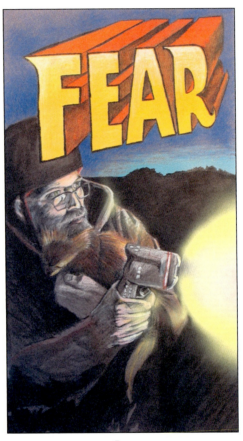

Fear
Illustration, 2016 • John Wood

Common Misperceptions about Dementia

Information can be empowering, so it's helpful to know that misperceptions about dementia exist. Misperceptions lead to stigmatizing attitudes and behaviors even among medical professionals. There are many misperceptions and misunderstandings about dementia.

One of the most common misperceptions is assuming everyone with dementia is living with advanced symptoms rather than realizing there are many types of dementia and that symptoms evolve over time. Misperceptions about dementia signal a lack of knowledge and understanding about the condition. The following are other common misperceptions about living with dementia —

Misperceptions	Facts
Everyone with dementia is the same.	Every person is a unique individual. People may have some similar dementia characteristics in the same way that people with brown hair share a similar characteristic.
People who have dementia are an empty shell.	This statement indicates a lack of understanding about dementia and fuels stigmatizing behaviors and practices. Being considered an empty shell is insulting and dehumanizing. People retain their personhood and individuality lifelong including with dementia.

Misperceptions	Facts
If you can speak for yourself, you don't have dementia.	This statement also signals lack of understanding about dementia and is insulting to those who are living with dementia. Besides the fact that there are many forms of dementia, people who have early to moderate dementia symptoms are very capable of speaking for themselves. People with advanced symptoms may lose some speaking capacity, but can continue to express themselves non-verbally.
If a person has trouble finding words, they also have trouble following a conversation.	Word retrieval and expressing language are controlled by a different part of the brain than receptive language ability.
People who have dementia cannot learn new things.	The brain has vast neural reserves and with stimulating triggers can form some new neural pathways resulting in learning.
You don't look like you have dementia.	There is not a particular 'look' to having dementia. Surprisingly, people with dementia hear this said to them.

When a person's network of family, friends, neighbors, health care providers, and local community members hold misperceptions about dementia, it can significantly and negatively impact everyday life. Richard Taylor was one of the first Americans living with early stage dementia symptoms to speak publicly and widely about his experience with stigma and how damaging stigma was to his emotional and social well-being.

> *"Family and friends questioned my ability to make everyday decisions — Can I be trusted to spend time alone with my granddaughter? Can I, should I handle my own money, answer the door... Friends stopping calling me and when I asked why was told, 'I don't know what to say.' I said, just say hello."*
>
> — Richard Taylor, PhD (Dupuis et al, 2011[5])

> *"Stigmas about dementia build massive barrier walls that can prevent us from having as fulfilled of a life as possible. Stigmas rob me of living my life to its fullest with dementia."*
>
> — Alexander (Sandy) Halperin

---

[5] Dupuis, S.L., Gillies, J., Carson, J., & White, C. (2011). *Moving beyond patient and client approaches: Mobilizing 'authentic partnerships' in dementia care, support and services.* Dementia, 0(0):1-26.

What You Can Do About Misperceptions

Until more people living with dementia publicly speak out and mass media and awareness campaigns catch up to accurate messaging, the onus of educating people about living with dementia mostly lies with each individual and family impacted. Dementia knows no boundaries and affects individuals of all income and education levels, across both genders, all ethnicities, and ages from as young as the 20s on up, although the majority of people affected are 60 and older.

Individuals who are living with dementia are terrific and powerful messengers. What better way to counter misperceptions than to calmly and in a friendly tone personally address the misperception directly! You don't want to put the person on the offensive as that generally shuts down any opportunity to change perceptions. Instead, addressing a misperception in a sincere and caring manner can be a compelling and effective way to change perceptions and educate people.

For example, sometimes when friends, extended family, neighbors and others know an individual has dementia, they will address the spouse or other family member and ask how the individual is doing even when the individual is right there. You can gently chuckle and say, "I'm right here and am happy to tell you how I'm doing. I'm doing great!" This friendly type of response can be the catalyst for a genuine, caring conversation about dementia and a chance to dispel misperceptions.

It's amazing how many doctors still address a family member instead of the individual who has dementia. If this happens to you, think of it as a chance to turn the tables and provide your doctor with an educational experience about LIVING WITH DEMENTIA and your interest in answering questions yourself.

Landscape
Ink Drawing, 2015 • Michael Crookes

Well-Being — A Newfound Friend

Up until now, you likely haven't given much thought to actively supporting your well-being. Generally, people think of well-being in terms of physical health, but well-being is broader and includes social, emotional, and spiritual dimensions as well as physical.

The "Preamble to the Constitution of the World Health Organization" defines well-being as — *A state of complete physical, mental and social well-being and not merely the absence of disease or infirmity.*

All dimensions of well-being are vital and important. Well-being is fundamental to the quality of one's life. Think about a time when you were at odds with someone important in your life and how it made you feel. You probably felt OK physically, but your emotional and social states may not have been OK affecting your overall sense of well-being. There are many ways well-being can be disrupted including not feeling personally safe or financially secure.

Discovering you have dementia and learning to live with it can significantly impact your well-being. As noted already but worth repeating, most people haven't given much thought to their well-being up until now. You must now. You will have to become intentional about maintaining your well-being. It will require active work on your part and does not naturally occur. The good news is the rewards far outweigh the effort!

It's important to stay socially, physically, emotionally and spiritually active lifelong. There is a natural tendency to let up on activity as one gets older and, more so, if facing a disabling condition. Letting up on activity is not a good choice because weakening any dimension of well-being has a cascading effect that weakens all dimensions. For instance, if a person withdraws socially, this affects the human need to connect with others and impacts them emotionally. The isolation also affects physical movement and limits the ability to keep one's body limber, flexible and strong.

A ROADMAP TO MAINTAINING YOUR WELL-BEING

- Keep your spirits up
- Develop and maintain a caring support network
- Limit stressful experiences as much as possible
- Seek fun and interesting experiences
- Expose yourself to laughter and humor
- Be open-minded to creative strategies
- Eat and drink healthily
- Learn good sleeping habits if you don't have them already
- Be physically active
- Increase opportunities to fill your spiritual soul

Keep Your Spirits Up

> *"Life is 10 percent what actually happens to us,
> and 90 percent how we react to it.
> We are in charge of our attitudes."*
>
> — Charles Swindoll[7]

One challenge of living with dementia is to stay positively and proactively oriented for your emotional well-being. It's easy to fall into a pattern of feeling sorry for yourself and having a negative outlook on life. Beware though as this action triggers other negative outcomes such as not being open to experiencing spirituality, engaging in meaningful and fun activities, having close connections with others, and being physically active — all of which have positive effects for well-being.

Negative attitudes are not helpful, but they will creep in at times. You may feel a lack of confidence when you cannot accomplish something you've done for decades. Although that's a normal reaction, find ways not to stay stuck in negative thoughts and instead find way to focus on what you can do and on being positive. Understand the change for what it is and then focus proactively such as writing about or recording your feelings and even laughing about it. Terry Berry laughs and thinks, "Great, one more thing I don't have to deal with anymore."

---

[7] Swindoll, C.R. (2006). *Great attitudes! 10 Choices for success in life.* Thomas Nelson Inc.

As noted above, life is 90 percent how we react to it. Attitudes are free, so you might as well choose a positive one. Everyone is different, so finding ways to stay positive may require some trial and error. Brian LeBlanc, who is living with Alzheimer's disease, has a wonderful mantra to keep himself positively focused — "I have Alzheimer's, but it doesn't have me."

Robert Bowles, who is living with Lewy Body dementia, created an acronym to remind him to stay positively oriented, **ASAP**.

A signifies *acceptance* of his dementia both in his head and heart

S stands for *socialization* to stay engaged with others

A means a keep a positive *attitude*

P stands for having a *purpose* to motivate and drive him

Suggestions

- Be of service to others. For many people, doing things that help others is uplifting. Some people may enjoy volunteering in their community while for others helpfulness can be as simple as picking up a neighbor's newspaper from the end of the driveway and placing it on the front stoop. It's hard to feel sorry for yourself when you are helping another.

- Listen to uplifting music.

- Understand that some things you do may work out differently than in the past. Instead of getting upset, go with a positive mindset — every day is a new adventure.

- Learn and try new things — take dancing lessons, try yoga, join a nature club, take up a sport, study a new language. Even if you are awful at the activity, you have found something new to laugh about!

- Consider volunteering somewhere you would find interesting. Mary Radnofsky, for instance, loves her volunteer work with children at the Smithsonian.

- Spend time with grandchildren and other young people. They have an open perspective about life that can be refreshing.

- If you find yourself with declining energy levels and more fatigue, choose the time of day to be more active when your energy levels are highest.

- Keep a pad of paper and pen handy so you can note something when you think of it instead of trying to remember what it was later.

Develop and Maintain a Caring Support Network

Even if you are a person hard-wired to 'do things myself,' now is the time to gracefully accept that you need support to maintain your independence for as long as possible. If you find it hard to ask for help, the good news is that asking for support gets easier each time! Surround yourself as much as possible with people who are positive and supportive. Be with people who make you feel good about yourself and see the upside and humor in life.

You'll find that most people want to help and be supportive but some may not know how. You may not know how they can support you either, in which case say so and ask them to work with you to figure it out. It's OK to be a bit awkward. Opening yourself up to others can be a special and bonding experience for you both. Let them know how much you appreciate and value their support. If someone tends to 'over help' or take over doing something for you, gently tell them you would rather do it yourself but with a little support from them.

A friend of John Wood told him she was happy that he is 'blooming where he is planted' beyond his dementia diagnosis.

> *"Knowing that the changes in my life were going to be inevitable, I decided to spend time where I could be accepted, challenged and grow. Thankfully, I have several different support groups, groups of friends and people with special interests. Being part of a caring network also means that I am able to make a positive impact by listening and speaking with others.*
>
> *— John Wood*

There is an online support network that receives rave reviews from people living with dementia, Dementia Mentors — www.dementiamentors.org. One of the services offered is to match newly diagnosed people with a person who has been living well with a similar type of dementia. People all over the world are involved, so you may find a new friend thousands of miles away! Dementia Mentors also offers Memory Cafes and online social gatherings.

Dementia Alliance International, a wonderful organization by and for people living with dementia, offers peer-to-peer support, online support groups and memory cafes among many other helpful supports — http://www.dementiaallianceinternational.org/events/cafe-le-brain/.

Limit Stressful Experiences

Stressful situations cause your central nervous system to respond by releasing stress hormones, such as adrenaline and cortisol, which prepare your body for action. Your heart beats faster, your muscles tighten, your blood pressure increases, your reaction time speeds up, and your senses become sharper and more focused. Experiencing occasional stress is OK. Regularly experiencing stress, however, isn't OK for your body as it can disrupt your immune system, increases the risk of heart attack and stroke, and negatively impacts your emotional, social and spiritual well-being.

Living with dementia symptoms can produce stress, so it's very important for you to establish ways to deal with stress in a healthy way. Sometimes stress cannot be avoided such as holiday events that can feel overwhelming. When there are stressful situations that cannot be avoided, find ways to limit your exposure to the stress. For example, during an event that includes a group of people make sure there is a quiet place to get away if you start to feel overstimulated. Bring earbuds to listen to soothing music or noise canceling headphones to create quiet and block out sound. Overstimulation can also affect other senses such as sight. Some individuals feel overwhelmed by too much color or pattern or too many items such as in a store or on a restaurant menu.

For some, a little stress can be motivating. Michael Ellenbogen and Brian LeBlanc, for example, both found experiencing some stress as a way of igniting their spirits and not giving up. Michael feels he is able to handle many things today because stress pushed him rather than made him feel like giving up.

When stress can't be avoided, have healthy ways to release the stress, such as take a walk, watch or listen to something funny, or ask your spouse or a friend for a back and shoulder massage. One person chops up a mass of vegetables to deal with stress and another likes to pull weeds from the garden. There are many ways to release stress, the key is finding what works for you.

Suggestions

- Avoid people who are unpleasant and negatively oriented. If you cannot avoid them altogether, you can limit the amount of time you spend with them. Set a time limit; knowing ahead of time, for instance, that you will only spend 20 minutes together can help you feel empowered and lessen stress.

- If you find something has become frustrating and stressful to do such as paying bills and managing your finances, decide whether YOU need to keep doing it. Perhaps your spouse or a trusted family member or friend would be willing to take this on for you. In return, you could reciprocate and do something special for them that doesn't cause you stress.

- Talk about how you are feeling with people to whom you are close. Verbalizing what causes you distress can be a benefit.

Seek Fun and Interesting Things to Do

There are endless ways to find fun and interesting things to do. Being socially engaged is one avenue of fun and doing interesting things. You don't have to join a party circuit! Being social can involve activities large and small, such as having coffee together, going to the movies or out for a meal; taking your dog for walks and seeing people along the way; shopping; and going on a trip.

Some people do well taking up a new hobby such as making pottery. For others, a new hobby may be a frustrating experience instead of something enjoyable. Paulan Gordon enjoys the former hobbies she didn't have time for while she was working.

Humans are hardwired for social engagement but the amount of engagement is individualized. Some people enjoy a lot of socializing while others just a little. What is important is not to become isolated. Social connections with people we are close to can be mood boosting, fulfilling, and enjoyable. These are valuable to the balance of your overall well-being. It is important for your well-being to invest the time to stay socially engaged and connected. Being socially active online with chat groups and helping others can be enjoyable and helpful if you live in a rural area or have limited access to transportation.

Throughout life your brain needs to be engaged in regular cognitive activity. While growth of neurons and synaptic connections weaken with age and dementia conditions, it does not stop altogether. Your brain needs stimulation to continue triggering generation of neurons and synaptic connections to form new neural pathways. Research shows there are important things to do that can make one's brain less vulnerable to the effects of disease and aging. There are lots of fun and interesting ways to stimulate and give your brain a workout.

Suggestions

- Go out and explore new things. Try a restaurant you've never been to before. Go for a walk with friends. One person, who lives near a city zoo, walks once a week with friends through the zoo. It is great exercise and they always see interesting things.

- See if there is a Memory Café near where you live. A state listing is available at http://www.memorycafedirectory.com/state-directories .

- Walk around your neighborhood regularly. It keeps you in touch with neighbors and activities going on around you.

- Enroll in a creative arts, writing or poetry course. You may find you have an inner artist.

- Take up a new hobby such as photography. If you can't manage some parts, it's a fun opportunity to do something with a spouse, friend, neighbor, former colleague, or grandchild.

- Consider becoming a Dementia Mentor and helping others — www.dementiamentors.org and www.demantiaallianceinternational.org .

Add Lots of Laughter to Your Life

Laughter is like a natural tonic. Laughter can actually improve your health! Laughter really is good medicine, here's how[8] —

- 🙂 Laughter relaxes your whole body. It relieves physical tension and stress.

- 🙂 Laughter boosts your immune system by decreasing stress hormones and increasing immune cells and infection-fighting antibodies.

- 🙂 Laughter triggers the release of endorphins, the body's natural feel-good chemicals. Endorphins promote an overall sense of well-being and can temporarily relieve pain.

- 🙂 Laughter protects the heart by improving the function of blood vessels and increasing blood flow that can protect you against a heart attack and other cardiovascular problems.

- 🙂 Laughter strengthens relationships. Sharing experiences that make you laugh with others is a bonding activity. Laughter can unite people during a stressful time and diffuse conflicts.

- 🙂 Laughter can help you release distressing emotions. You can't feel anxious, sad, angry or frustrated when you're laughing.

- 🙂 Laughter is free and abundantly available!

---

[8] HelpGuide, a collaboration with Harvard Health — http://www.helpguide.org/articles/emotional-health/laughter-is-the-best-medicine.htm .

Suggestions

- Smile, it's the beginning of laughter.

- Spend time with people who incorporate humor and laughter into their daily lives. It's contagious.

- Watch a funny movie, TV show or family videos.

- Laugh at yourself. You'll take yourself less seriously when you can laugh at your mistakes and antics.

- Recount funny things that happened to you with others. Not only will it make you laugh again, but likely will spur others to share their funny experiences.

- Watch live animal video cams. It's hard not to chuckle at puppies or kittens crawling over each other.

Betty
Acrylic on canvas, 2015 • Anonymous

Increase Opportunities to Fill Your Spiritual Soul

Spirituality is intended here in a broad context and not as 'religion.' While participation in a religion may be a meaningful way to experience spirituality for some, there are many other ways such as walks in nature, service to others, enjoying the arts and music, seeing a sunset, meditation, and spending time in a special place. Spirituality does not have to involve a deity or a higher power. It is feeling deeply connected to something larger than ourselves such as nature or being of service to others.

Taking the time to intentionally fill your spiritual soul is especially important when managing day to day living with a health condition such as dementia. It can be easy to let the challenges overwhelm and overcome you if you don't have a process for experiencing relief. Feeling the sun on your skin or sand on your feet, seeing something touching such as a rainbow, sunset or field of flowers, listening to moving music or just going for a walk can be transformative experiences.

Suggestions

- ♦ Take time to 'turn off your mind' and just relax. John Wood finds it difficult to just sit to be quiet, so instead he goes to a quiet place that is meaningful to him — a museum.

- ♦ Laurie Scherrer has learned to appreciate the sounds and beauty of the many birds that visit her backyard. To her delight, she has found her backyard is inviting to an assortment of birds, including many rare and migrating birds. She especially enjoys watching "ALBI," an albino bird of unknown species, fly across her windows — just like an angel. The birds' melodies and elegance helps calm her mind.

♦ Consider taking a meditation class.

♦ Listen to music that moves you. For some it can be an opera, others spiritual music or jazz. The type of music doesn't matter so long as you find it soothing and moving.

Fireflies
Mixed media • Anne Garavaglia

Be Open-Minded to Creative Strategies

Sometimes, caring, well-meaning family members and friends take over doing things for you instead of helping support you so you can continue doing things for yourself. When others take over doing things for you, it opens up the potential for learned helplessness and disengagement. Kate Swaffer coined a term, Prescribed Disengagement®, for the disabling effect experienced when others inadvertently take choice and control away from a person living with dementia[9].

> *"(Prescribed Disengagement®) sets up people with dementia to believe there is no hope, there are no strategies to manage the symptoms of dementia, and more importantly, that it's not worthwhile trying to find any."*
>
> — Kate Swaffer, with permission, p. 160

Richard Taylor described his preference, "I need you to help enable me, not disable me." Compensatory strategies that help enable individuals with dementia to continue being proactive and do things for themselves are beneficial for all dimensions of well-being. Compensatory strategies, workarounds, and roundabouts are all terms used to describe needing to be creative to find alternative, supportive ways to accomplish things that now present some challenges.

---

[9] Swaffer, K. (2016). *What the hell happened to my brain? Living beyond dementia.* Jessica Kingsley, London, p. 160.

For instance, Terry Berry lives alone with her dog but close by to two daughters. Her daughters noticed the dog was losing weight and realized their mother was forgetting to feed the dog. Their ingenious strategy was to use an empty mayonnaise jar and mark the outside with permanent ink to show "full" and "half" levels. Every morning Terry fills the jar to the "full" level with dog food and gives the dog half of the jar of food. She leaves the jar on the kitchen counter to serve as a reminder and in the evening gives the other half of the food in the jar to her dog for dinner. She leaves the empty jar on the counter so the next morning she'll see the empty jar and be reminded to fill it up and start the cycle again.

There are innumerable creative strategies to help you stay independent and doing things for yourself. All you need is to be open-minded, think creatively, and, at times, ask others for ideas about ways you can continue to manage an activity. Being resourceful and staying proactive is life affirming.

Suggestions

- ♦ Place items such as keys and eye glasses in a consistent location so you don't have to remember where you last placed them. Give yourself a visual cue such as a picture of keys and eye glasses in the consistent location so the picture can serve as a friendly reminder to place the item back when not in use.

- ♦ Large crowds and lots of noise may become overwhelming. If so, you can grocery shop or go shopping at the mall when it's least busy — usually when the stores first opens or later in the evening. Go early in the morning if you're a morning person or later in the evening if you're more energized at that time of day.

- Many people living with dementia find technologies helpful. There are many different varieties of smartphone app memory aides for reminders, managing your schedule and for daily routines. There are technology systems for managing medications, and translation software to convert written material to a spoken form for individuals experiencing visual impairments. These are just a few of the many ways technologies can be supportive.

- Brian LeBlanc asked his neurologist to write a statement noting that he has Alzheimer's disease and needs to be accommodated to board the airplane in advance of other passengers. This allows Brian to find his seat and get settled before the plane fills up with people and noise.

Blossom Tree
Mixed media, 2015–2016
Collaborative – Residents of Springwood Residential Home

Eat and Drink Healthily

Eating a healthy, balanced diet is important for your physical health. Not eating enough can lead to weight loss and other problems such as vulnerability to infection, reduced muscle strength and fatigue. Drinking and staying well hydrated is just as important as healthy eating. Dehydration can lead to headaches, constipation, and urinary tract infections — all of which can worsen symptoms of dementia. Heart healthy diets have been found to be good for the brain. You should choose foods low in salt, low in added sugar and artificial sweeteners, and low in unhealthy fats. It's never too late to start to eat healthily.

The following is a healthy balanced diet recommended by the U.S. National Institute on Aging for people 50 years of age and older[10]:

- Fruits—1 ½ to 2 ½ cups
  What is the same as a half cup of cut-up fruit?
  A fresh 2-inch peach or 16 grapes.

- Vegetables—2 to 3 ½ cups
  Two cups of uncooked leafy vegetables is the same as a cup of cut-up vegetables.

- Grains—5 to 10 ounces
  A small bagel, a slice of whole grain bread, a cup of flaked ready-to-eat cereal, or a half cup of cooked rice or pasta is the same as an ounce of grains.

---

[10] Retrieved online on December 8, 2016 at (https://www.nia.nih.gov/health/publication/healthy-eating-after-50).

- Protein foods—5 to 7 ounces
  One egg, one fourth cup of cooked beans or tofu, a half ounce of nuts or seeds, or 1 tablespoon of peanut butter is the same as an ounce of meat, fish or poultry.

- Dairy foods—3 cups of fat-free or low-fat milk
  One cup of plain yogurt or 1½ to 2 ounces of cheese is the same as 1 cup of mild. One cup of cottage cheese is the same as a half cup of milk.

- Oils—5 to 8 teaspoons
  Foods such as olives, nuts, and avocados have healthy oil in them.

- Solid fats and added sugars (SoFAS) and sodium (salt)—keep the amount of SoFAS and sodium small
  If you eat too many foods containing SoFAS, you will not have enough calories left for the more nutritious foods you should be eating.

- Your doctor may want you to follow a special diet because you have a health problem like heart disease or diabetes. Or, you might have been told to avoid eating some foods because they can change how well your medications work. Talk with your doctor or a registered dietitian, a nutrition specialist, about foods you can eat instead.

- Stay away from "empty calories." These are foods and drinks with a lot of calories but not many nutrients — for example, chips, cookies, soda, and alcohol.

- Drink 64 ounces of fluid such as water daily.

For additional information about healthy, balanced diets, the following is the link to the "Dietary Guidelines for Americans 2015 — 2020". https://health.gov/dietaryguidelines/2015/resources/2015-2020_Dietary_Guidelines.pdf

Landscape
Acrylic on canvas, 2015 • Gordon

Learn Good Sleeping Habits If You Don't Have Them Already

Sleep plays a vital role in maintaining your physical health. Most of us are unaware of all the vital functions sleep contributes to our overall health. Getting sufficient, quality sleep helps you function during wakeful hours. Because bodies need sufficient sleep for healing and repair of heart and blood vessels, a deficiency of sleep increases the risk of heart disease, kidney disease, high blood pressure, diabetes and stroke. Several nights of losing even 1 to 2 hours of sleep can affect your body's ability to function normally.

Getting sufficient, quality sleep helps your body maintain a healthy balance of the hormones that make you feel hungry or full. When you do not get enough sleep, your level of the hunger hormone (ghrelin) goes up and the level of the full hormone (leptin) goes down resulting in you feeling hungrier and less full than normal. Your immune system, vital to your body's defense against infections, also relies on sleep to stay healthy. Ongoing sleep deficiency can change the way your immune system responds to fighting infections[11].

Many people are not even aware they aren't getting sufficient sleep. There is some genetic variation, but most adults need between 7.5 to 8.5 hours during each 24-hour period. People who have dementia can experience a change in their sleep patterns. You may feel sleepy during the day, have difficulty sleeping for long periods at night and experience

---

[11] Retrieved online on November 2, 2016 at https://www.nhlbi.nih.gov/health/health-topics/topics/sdd/why.

a lighter, less restful night sleep. Because so many people are unaware of the important role sleep plays for physical well-being, the following are some helpful sleep recommendations.

Suggestions

♦ Maintain your regular sleep routines as best you can. If you are used to falling asleep to music, listen to your favorite music. If you typically get up early, keep getting up at the same time. Keep the same getting-ready-for-bed routine as well.

♦ Stay ACTIVE during the day. Being exposed to bright light or sunlight controls melatonin levels and sends the brain a message to be awake.

♦ If you want to nap during the day, the best time is early afternoon so you are tired and ready to sleep at night.

♦ At night, keep your sleep area dark and comfortable with a small night light to help you find your way safely if you get up during the night.

Twilight
Mixed Media • Anne Garavaglia

Be Physically Active

Physical activity is vitally important for staying limber and flexible, maintaining balance and muscle strength and staying proactive about your life. While some people may enjoy regimented calisthenics, others may bristle at the notion. Choose whatever sustained physical activity you will enjoy doing. Dancing, for example, is fun and good for flexibility, balance, muscle strength and sustained activity. Dancing has also been found to be beneficial for cognitive stimulation!

Physical activities such as tai chi or yoga are also good forms of relaxation. If you experience balance challenges, you may need to tailor yoga or tai chi activities to work for you. There are many ways to stay physically active. Some physical activities also accomplish tasks such as raking, sweeping, scrubbing the floor, and cleaning windows, while other activities are just for fun like walking, biking, and jogging.

Being physically active doesn't just happen. You need to plan activities and be mindful of your level of activity.

Suggestions

- ♦ Besides being beneficial for your health, getting involved in physical activities such as yoga, boxing, walleyball, and swimming to name a few increases your social interactions.

- ♦ Doing a physical activity with another person can give you something to look forward to and provide companionship. It's also a good way to stay connected to someone you care about.

♦ Be open minded. One individual was never a fan of square dancing until he got dragged to an event. He loves it and it's now something he looks forward to weekly.

Cat as Trophy
Acrylic on canvas, 2016 • Doris Ann Brucker

Finding Silver Linings in Your Unexpected Journey

Clearly living with dementia is not all wine and roses. It's not all doom and gloom either. Jim Taylor sums up his feelings about his wife's dementia —

> "I know this sounds really strange, but I don't think Geri and I have had a better period in our marriage than right now. We're much more dedicated to be with each other. The disease has brought that. Yes, it's short-lived and there's a sadness to that, but we're not focused on that."
>
> — Kleinfield, 2016[12]

> "Where do I go from here? I have no idea. My motto is, "one day at time." I will attempt to stay positive, passionate, energetic, help others and advocate for dementia awareness and education. We all need purpose."
>
> — Robert Bowles

---

[12] Infield, N.R. (April 30, 2016). *Fraying at the Edges*. The New York Times.

ATTACHMENT I.

What is Dementia?

Dementia is an umbrella term that describes a group of cognitive symptoms such as difficulty remembering, concentrating or problem-solving; shortened attention span, confusion with location or the passage of time; impairments in judgment, word-finding, and learning; and changes in behavior and personality. These types of cognitive changes are caused by damage of the brain's nerve cells, or neurons. Alzheimer's disease is the most common form of dementia. Other common forms of dementia are vascular dementia, caused by stroke or blockage of blood supply to the brain; Lewy Body dementia; frontotemporal dementia; and mixed dementia, a combination of several forms of dementia.

The symptoms and the progression of dementia vary depending on the type of disease causing it. Some forms of dementia progress slowly over years, while vascular dementia may result in sudden loss of some cognitive functions. Although dementia is more common in older adults, it is not a part of normal aging. People younger than age 65 can also be affected. In the United States, it is estimated that 200,000 people younger than age 65 have dementia.

Often people do not recognize the changes in themselves; others who have regular contact with them notice the changes first. The adage, "Ignorance is bliss," may be useful for some things in life, but it is not an advisable strategy for one's health. If a person is exhibiting some symptoms, it is important to be assessed by a physician. The symptoms may be a result of other conditions that are treatable such as thyroid problems, vitamin B-12 deficiency, reaction to a medication, infections, and depression among other possibilities.

Because dementia is a condition that generally evolves over many years, some people refer to 'stages' to identify the levels of symptoms being experienced. The terms 'early,' 'mid,' and 'advanced' dementia are only for general guidance. Every person's dementia journey is unique and personal.

Garden Tree
Encaustic, 2016 • Carol Ambrogio Wood

ATTACHMENT II.

Assessment & Diagnosis of Dementia

Early on you might not even realize you have a cognitive problem. The symptoms are often more obvious to a family member or friend in day-to-day living. Not recognizing symptoms yourself can impact your ability to get assessed. It is helpful to bring a close family member or friend along on the physician visit so she or he can provide observations that you may not be aware of. It is important to see a physician who has expertise in this area and not, for example, one's urologist, cardiologist, or gynecologist. While many internal medicine and family practice physicians can accurately diagnosis types of dementia, not all have the specific training or experience. Some forms of dementia such as frontotemporal, Lewy Body and posterior cortical atrophy can be difficult to diagnosis unless the physician has specialized expertise. For instance, many people who have Parkinson's disease also have Lewy Body dementia. It is recommended to see a geriatrician (physician with expertise in older adult health care) or a neurologist with expertise in dementia.

Considerations

There may be some bumps in the road to getting assessed and diagnosed. The following are some suggested helpful steps:

- ♦ Keep a list handy to write down changes you may be experiencing so you won't have to rely on memory when speaking to a physician. Also, invite others close to you to write down changes they may observe.

♦ Bring someone you trust to see the physician with you. It can be hard to take in all the information yourself. Ask your companion to take notes during the physician visit so that you both can review and discuss the information later.

♦ You may feel uncertain about what is involved in a dementia assessment. If so, call the physician's office and ask them to describe the assessment process, where everything will take place, and how much time is involved.

♦ If you and your companion don't feel confident or comfortable with the assessment and/or diagnosis, seek a second medical opinion.

♦ Ask your physician if she or he has the training and experience to help you manage this condition or if she or he could recommend a specialist.

ATTACHMENT III.

Your Brain

It helps to have a general understanding of your brain to understand how changes in your brain can affect various cognitive functions.

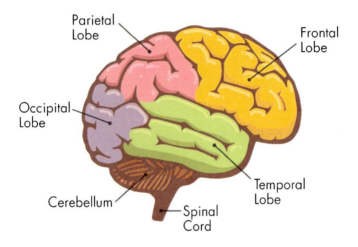

Areas of the brain

The PET scan images on the next page show typical patterns of brain activity for four different cognitive functions. The red color indicates brain areas where activity is the highest. Yellow represents a decrease of brain activity, then green, and then to blue-violet that indicates the lowest level of activity. Different parts of the brain control different functions. Reading words uses a different part of the brain, for instance, than thinking about words such as what day it is.

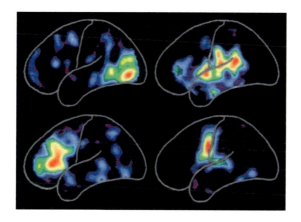

http://www.alz.org/braintour/signal_activity.asp

1. **Reading words** (top left image)
2. **Hearing words** (top right image)
3. **Thinking about words** (lower left image)
4. **Saying words** (lower right image)

Typically the brain processes information and actions in milliseconds. If your brain has neural damage, it is easy to understand how information and actions can malfunction. Below is a breakdown of steps for a bill paying activity as an illustration of the many steps the brain needs to process. Some of the steps people may not even be consciously aware they need to do.

1. Idea to pay bills

    - Think about when the bills are due and if this is the right time of the month to pay them

    - Check to see if the bank account has the funds to cover the bills

2. Gather materials to pay bills

   - Think sequentially about all the items that are needed to pay bills
   - Remember where these items are stored and get them

3. Pay bills

   - Identify how much is due for each bill
   - Pay the amount (online process or write a check)
   - If paying by check, put bill and payment in envelope; you may need to address the envelope; add postage stamp
   - Know when paying the bill is completed

4. File the bills and put away other items such as the checkbook, stamps, etc.

5. Mail the bills

   - Check that bills have addressed envelopes
   - Add a stamp to each bill and return address information
   - Place bills in mailbox

6. Stop thinking about the bills and move on to something else

Scientists once thought that the brain stopped producing new neurons early in life. They now know this isn't so; the brain continues to produce new neurons, called neurogenesis, throughout life although less rapidly as people age. There is emerging science about neuroplasticity which is the brain's ability to reorganize itself by forming new neural connections. These new neural connections can enable the brain to compensate for impaired areas. There is still much to be learned about neuroplasticity and how to promote and enhance new neural connections to compensate for cognitive impairments.

There is an urban myth that humans only use a small part of their brain. The truth is we use all of our brain. Neurons are continually discharging nerve impulses. If the nerve impulses are not actively used, however, they will lose their connectivity. Neurons need to be stimulated to stay active (Hood, 2012[13]). This is valuable information for you to know as encouragement to stay mentally active and engaged. Stay involved, be productive and do things that are meaningful to you — it's good for your brain!

---

[12] Hood, B. (2012). Why you don't use only a fraction of your brain. Psychology Today. Downloaded online December 8, 2016 at https://psychologytoday.com/blog/the-self-illusion/201206/why-you-dont-use-only-use-fraction-your-brain.

ATTACHMENT IV.
Managing Dementia Symptoms: Medications

This booklet does not provide medical information. It is about LIVING life with dementia, so this may include your decision to take medication. While there are no treatments for dementia, some medications have been found to temporarily improve some of the symptoms associated with dementia for some people. The decision to take any medication should be discussed with your physician. The following are the medications approved by the U.S. Food and Drug Administration for dementia —

- Cholinesterase inhibitors — These medications — including donepezil (Aricept), rivastigmine (Exelon) and galantamine (Razadyne) — work by boosting levels of a chemical involved in memory and judgment. Although primarily used to treat Alzheimer's disease, cholinesterase inhibitors might also be prescribed for other dementias, including vascular dementia, Parkinson's disease dementia and Lewy body dementia. Side effects can include nausea, vomiting and diarrhea.

- Memantine — (Namenda) works by regulating the activity of glutamate, another chemical messenger involved in brain functions, such as learning and memory. In some cases, Memantine is prescribed with a cholinesterase inhibitor. A common side effect of Memantine is dizziness.

- Other medications. You may find that you experience other symptoms or conditions besides cognitive changes, such as anxiety, depression, sleep disturbances or agitation. If this is the case, you should speak with your doctor the need for any prescribed medications to treat any other conditions.

Zieminski's Grocery
Photograph • Ronald T. Konopka

Resources

There are many helpful resources and supports available that have been curated by individuals living with symptoms of dementia. The Dementia Action Alliance's website maintains an up-to-date resource list of helpful blogs, books, Facebook pages, publications, research centers, videos, and websites from around the world. Please visit http://daanow.org/resource-center/.

Dementia Mentors — www.dementiamentors.org — and Dementia Alliance International — www.dementiaallianceinternational.org — are two excellent sources to connect with others living with symptoms of dementia. They both offer online chats and social gatherings. Dementia Mentors offers to match newly diagnosed people with a person who has been living well with a similar type of dementia.

Hand
Acrylic on canvas, 2016 • Maria Nightingale

About the Dementia Action Alliance

DAA's Vision

The Dementia Action Alliance (DAA) is a diverse coalition of passionate people creating a better society now for individuals to LIVE with dementia. DAA envisions a society where dementia symptoms are better understood and accommodated as a disability, and individuals and families living with dementia are fully included and supported.

DAA's Mission

The Dementia Action Alliance is a diverse coalition of passionate people creating a better society now for individuals to LIVE with dementia.

DAA's Goals

- WORK directly with individuals who have dementia to learn from and amplify their first person perspectives about dementia.
- EDUCATE the public about living with dementia to raise awareness and increase understanding. Lack of information fosters misperceptions and stigmatizing behaviors toward individuals with dementia.
- COLLABORATE with the diverse dementia community for collective impact to advocate for public policies, practices, and research that optimizes the well-being of people living with dementia.
- CREATE, curate and post free person-centered dementia support resource materials online.
- SUSTAIN the operation of the Dementia Action Alliance.

The Artwork

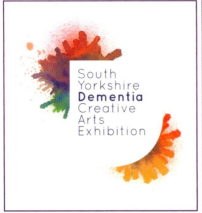 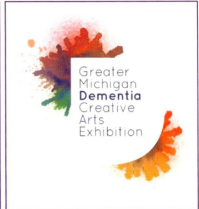

The inspirational artwork throughout the booklet are part of a unique collaborative effort of David Reid, Senior University Teacher at the University of Sheffield, England and John Wood, a visual artist living with dementia in Detroit, Michigan. David, as part of The University of Sheffield's 'Engaged Learning,' strand of community partnership work, established the South Yorkshire Dementia Creative Arts Exhibition eight years ago. The exhibition features works by individuals who are living with dementia and their care partners to enhance public understanding about life lived with dementia. The exhibition has inspired others and now travels internationally.

John was one of the people inspired by the South Yorkshire Exhibition. Working with others, John created the Greater Michigan Dementia Creative Arts Exhibition in 2015. "Our goal for the exhibition is to be inclusive to all persons involved in a dementia diagnosis. Hopefully the artworks can remove the stigma related to lives affected by dementia."

Meet the Artists

Anonymous – UK **"Betty"** [care partner]

"'Betty' was produced by my former colleague. It is a portrait of Betty, who has been coming to The Wellbeing Centre for years. She wanted to catch her joyful yet stern character through the use of colour & black outlines." ~ Natasha Wilson

Linda Birtles – UK **"Leaves"** [person with dementia]

I took my camera and paints on an early autumn visit to Nottingham Centre Parcs. Whilst wandering through the woodland, I was moved by the changing colours of a beech tree in the fading light as winter approached.

Doris Ann Brucker – USA **"Cat as Trophy"** [care partner]

"The artist was inspired by an antique print hung in her children's bedroom. The five senses: seeing, hearing, smelling, touching, and tasting, are demonstrated cleverly by five inquisitive cats."

Collaborative - UK **"Blossom Tree"** [care partner]

"This artwork was created by the residents of SheffCare's Springwood Residential Home in Sheffield during our weekly craft sessions. The residents used their hand prints to create branches of our Blossom Tree giving this piece a sense of identity." ~ Sarah Simmonite, Arts Coordinator

Michael Crookes – UK **"Landscape"** [person with dementia]

"Michael uses all types of materials in his artwork, except oils. He currently runs art classes in his local area of Crookes. Michael is inspired by things he sees around him and always carries a pencil and paper with him — ready to sketch."

Anne Garavaglia – USA **"Fireflies"** [care partner]

"Fireflies" speaks to the sudden light from a firefly appearing before them. The light lasts but a moment. It is suspended time when the viewer tries to follow the mysterious insect's last trajectory in hope of seeing its glow again — much like trying to follow a train of thought which suddenly becomes elusive.

Anne Garavaglia – USA **"Twilight"** [care partner]

"Twilight" speaks to the redemption of memory, the hope of enlightenment, the strength of recall.

Gordon – UK **"Landscape"** [person with dementia]

"The landscape was produced by Gordon, a member of the AgeUK Wellbeing Centre in Sheffield. He used an image found online to interpret and inform the landscape he created. Gordon spent several weeks perfecting the image." ~ Natasha Wilson, care partner

Ronald T. Konopka – USA **"Zieminski's Grocery"**
[person with dementia]

"Mr. Konopka's commitment to detail displayed in his artwork allows the viewer to appreciate the thoughtful care and affection he has for his memories. His work depicts his family history through the grocery stores and florist shop of his grandparents and parents."